YOUR KNOWLEDGE HAS VALUE

- We will publish your bachelor's and
  master's thesis, essays and papers

- Your own eBook and book -
  sold worldwide in all relevant shops

- Earn money with each sale

Upload your text at www.GRIN.com
and publish for free

Imprint:

Copyright © 2017 GRIN Verlag, Open Publishing GmbH
Print and binding: Books on Demand GmbH, Norderstedt Germany
ISBN: 9783668461284

This book at GRIN:

http://www.grin.com/en/e-book/367200/application-of-capillary-electrophoresis-in-
quantification-of-toxins-in

Oladimeji Adewusi

Application of Capillary Electrophoresis in quantification of toxins in food

GRIN Publishing

Application of Capillary Electrophoresis in quantification of toxins in food

Oladimeji Adewusi

Food Analysis Term paper, University of British Columbia.

Abstract

Capillary Electrophoresis is a separation technique that presents scientists with lots of advantages over other separation methods. In this review, the various modes of capillary electrophoresis and different detectors coupled to it were discussed. Toxins commonly found in food is presented as well, with emphasis on natural toxins such as mycotoxins, bacterial toxins, and paralytic shellfish toxins. Finally, applications of capillary electrophoresis for toxin analysis in different studies were summarized.

Keywords: Capillary electrophoresis, toxins, detectors, natural toxins

Content

Introduction

Food analysis may be viewed as a two-step activity involving separation stage and detection step. The effectiveness of food analysis depends on the accuracy and precision of both steps. Food ingredients and food products are analyzed for several reasons including determination of nutrient composition, evaluation of quality attributes, and detection of hazardous substances such as toxins.

Certain substances of natural origin in food pose significant health risk when ingested. These substances are known as toxins. Processing and handling of foods can also result in accumulation of toxins in foods thus toxins may range from microbial sources such as mycotoxins, neurotoxins in shell fish to toxins that accumulates during growth such as glycoalkaloids in potatoes and in processing such as acrylamide.[1] Rapid detection and quantification of toxins in food is important because of the health hazard associated with it. Methods that have been used in detection of food toxins have been reported and they include biological assays such as ELISA, bioluminescence assay and protein phosphatase inhibition assay, high performance capillary electrophoresis, chromatographic methods like HPLC and GC/FID.[2] Amongst these analytical techniques, capillary electrophoresis can be regarded as a more efficient and fast method of toxins determination when coupled to a detection system.[3]

Capillary electrophoresis was introduced in the 1980s and started gaining popularity recently because of its versatility as a separation method. It is used in the analysis of multiple components of food, including ionic and non-ionic components of food.[3]

Several advantages associated with capillary electrophoresis over other separation methods include versatility of application. This implies that a wide range of analytes such large biomolecules like DNA, inorganic compounds and organic molecules can be separated using this method. Also, the possibility of employing different modes of separation on the same piece of equipment as well as its ability to interface with different detection systems have increased the use of capillary electrophoresis in analytical chemistry. Other advantages such as minimal use of sample, solvent and ruggedness of capillary electrophoresis have been reported.[4, 5]

Although capillary electrophoresis presents analyst with high separation efficiency and fast separation, its main drawback is its poor sensitivity due to its low dimension. There will be a compromise with the resolution of CE if volume of injected sample is increased to improve sensitivity. Other methods of improving sensitivity of capillary electrophoresis which largely involve the pre-concentration of samples prior to separation with CE have been reported. These include solid phase extraction, sweeping and field amplified stacking.[6]

This articles gives a comprehensive review of the application of capillary electrophoresis in the separation and quantification of various toxins in foods.

Capillary electrophoresis

Capillary electrophoresis is a collective term for methods used in separation of molecules through narrow tubes when electric field is applied. The difference in migration of the charged molecules results in separation of analytes in a sample.[3] Based on this, capillary electrophoresis is largely used for the separation of chemical compounds with similar structure.[7]

Before the introduction of capillary electrophoresis, the traditional electrophoretic method was first described by Tiselius in 1937 and it was used to separate protein fractions in a sample placed in buffer solution with applied electric voltage. However, the separation was poor due to thermal convection effect. To mitigate this, slab gels were introduced to the electrophoretic system leading to the invention of gel electrophoresis with relatively higher separation efficiency when compared to the traditional electrophoresis but lower separation efficiency when compared to HPLC and GC. To improve the separation efficiency and analysis time of gel electrophoresis, narrow tubes were introduced to the electrophoresis system for rapid heat dispersion, thus capillary electrophoresis can be viewed as an improvement of the slab gel electrophoresis system.[3]

Different modes of separation are employed in capillary electrophoresis depending on the type of analytes. Capillary zone electrophoresis (CZE) is a separation mode that involves migration of charged molecules (positive and negative) to the terminal end of a fused silica capillary filled with electrolyte. This is caused by the action of electro-osmotic flow. The velocities at which the ionic species migrate within an electrophoretic system differs and can be calculated by mathematical relation. This velocity of migrating species is largely dependent on electrophoretic mobility and applied electric field. This implies that charged species tend to move faster at higher electric field. Also, the choice of buffer affects the efficiency of separation because pH and composition of buffer affect mobility of charged species.[5] The various parts of capillary electrophoresis is shown in figure 1.

Although the separation efficiency of CZE is high, however, it has the drawback not being able to separate neutral molecules. To overcome this, Micellar electro kinetic chromatography

(MEKC) was developed. Chromatography principle is employed for the separation of neutral molecules using a pseudo-stationary phase by adding surfactant such as sodium dodecyl sulfate to the background electrolyte thus MEKC is used for separating both neutral and charged molecules. Other surfactants that can be used include bile salts and quaternary ammonium salts. [3,5]

Capillary electrochromatography is a less common CE mode although high separation efficiency is achievable through this mode. It is unique because of the striking similarities it shares with a regular chromatographic method in that it uses a silica packed column as stationary phase with electro-osmosis being responsible for the mobile phase flow. [8]

Capillary Isotachorphoresis (CITP) is another peculiar CE separation mode in which ionic compounds are distributed in discrete zones between a leading electrolyte and terminating electrolyte with different flow velocity while the discrete zones move at the same velocity. Steps are used to denote analytes concentrations on an isotachopherogram in capillary isotachorphoresis mode as oppose to CZE mode where analytes intensities are represented by peaks on electropherogram. [3,8]

The possibility of separating molecules based on their Isoelectric point is the principle employed by Capillary Isoelectric focusing mode of CE.it is based on the principle that when a capillary is filled with protein sample along with a solution that can create pH gradient. The instrumental set up is such that the anodic end of the capillary is inserted into an anolyte while at the same time the cathodic end is immersed in catholyte. The protein sample becomes separated after migration to a zone in the capillary where it has neutral pH. [8]

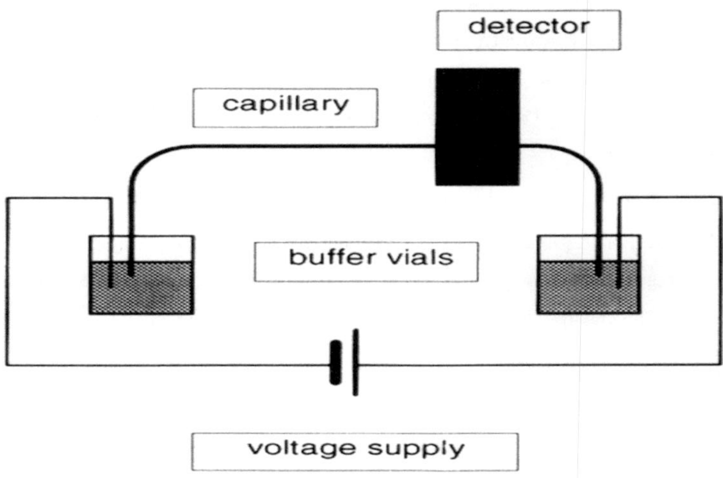

Figure 1. Diagrammatic representation of capillary electrophoresis

Methods of detection

Since CE is essentially a separation method, it cannot be used to determine the concentration of an analytes, thus it is usually coupled to a detector that can quantify the amount of analyte in a sample. CE has the drawback of tiny peak volumes because of the very small amount of sample injected in to the capillary, thus a sensitive detection method of detection is required to give accurate concentration of analyte.[4]

Several methods of detection are available to CE but UV-VIS detector is the most commonly used. These techniques include laser induced fluorescence, electrochemiluminiscence, conductometric detection, refractive index detection, and amperometric detection.

UV-Vis detector can be considered as in column detector because they are coupled directly to the CE equipment and detection is done by creating a window space on the column by scraping off the polyimide coating on a section of the CE. The observed drawback of UV-vis detector is obvious because of low dimension of capillary. This results in low concentration sensitivity as well as short linear range. The use of z-shaped capillary and bubble cells have been suggested as means of improving the path length in UV-vis detector[6]

Laser induced fluorescence (LIF) is a very sensitive in column detector coupled to CE, with advantage of selectivity. High energy excitation is possible with Laser induced fluorescence leading to high radiation output. This is the reason behind the sensitivity of LIF. However, the instrumentation is usually complex, expensive and it requires appropriate sources of laser. [2,6]

Micellar electrokinetic chromatography (MEKC) coupled to fluorescence detection has been used successfully to detect paralytic shellfish poisoning toxins in shellfish. In the same vein, detection of tetramine, a toxin found in whelks was detected with Electron Spray Ionization-Mass Spectroscopy (ESI-MS) after separation with capillary electrophoresis[7]

Electrochemiluminiscence detector have also been coupled to CE for detection of analytes. However, they are mostly applicable in analytes containing tertiary amine group. At the electrode surface, analytes in the CE effluent reacts with $Ru(bpy)_3^{2+}$ and give rise to an electropherogram based on the characteristics of the analytes. [8] Capillary electrophoresis coupled to electrochemiluminiscence detector is shown in figure 2.

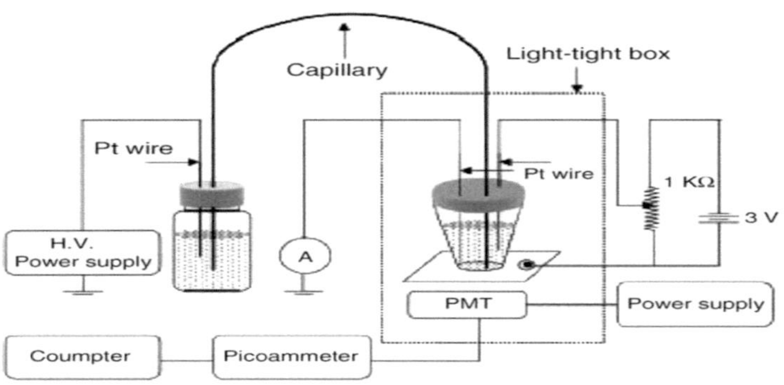

Figure 2. Capillary electrophoresis coupled to electrochemiluminiscence detector

Equipment set up

The instrumental set up of capillary electrophoresis is made up of a silica packed capillary, voltage supply, detector for on column detection, injection system, electrode jars (buffer), electrodes. Capillary is one of the most important part of a CE because its attributes affect the overall performance of the equipment and separation and detection takes place on it. Capillary for CE comes in various length (20 – 100 cm) and different internal diameter (20-100μm). For a good performance, capillary should be chemically and physically resistant. Other desirable characteristics include flexibility (achieved by external coating with polyimide), low cost, transparency (needed for UV detection).

Sample injection in CE can be achieved either by controlling injection pressure (hydrodynamic principle) or voltage (electro kinetic principle). It is important to note that because of low dimensions of capillary, small volume of sample is required to be injected into the capillary for a good separation efficiency.

High voltage supply delivers up to 35W to separate analytes based on their charge to mass ratio. CE separation are preferably done at a constant voltage supply however constant current can also be used.[9,10,11]

Toxins

Toxins refer to poisonous substances produced by living cells as metabolites. The term has also been associated with other form of poisonous substances not of biological origin. For instance, toxic materials produced arising food processing may be conveniently referred to as toxicants and not as toxins. Foods have been consumed mainly because of the nutrients however food materials may also contain toxic principles such as cyanogenic glucosides in cassava, solanine in potatoes, heavy metals and biogenic amines in fish and aflatoxins in peanut. Toxins produced in plant food materials are generally known as phytotoxins [12,13]

Natural toxins are considered significant biological and chemical hazard; therefore, it is a major concern in food safety. Natural toxins include marine toxins, bacterial toxins, mycotoxins, phytotoxins, mycotoxins as well as biogenic amines in fish.[12]

Mycotoxins

Mycotoxins are toxic substances produced by fungi as secondary metabolite. They are a group of small molecular weight chemical compounds with ability to cause harm to humans thus constitute a significant chemical hazard in food chain. Usually, mycotoxins are produced by saprophytic molds growing on crops on field or harvested food materials in storage. The major genera known for producing mycotoxins amongst the saprophytic molds are Fusarium, Aspergillus, and Penicillium. Mycotoxins are usually found in cereals because when not stored properly, mold growth is aided. They may be found in wide range of food products including beer made from mildewed barley, wine, milk, egg as well as food materials such as coffee, cocoa, spices, dried fruits. aflatoxins, ochra-toxin A, fumonisins, deoxynivalenol, T-2 toxin and zearalenone are the most common mycotoxins found in food and feed. [14,15]

Aflatoxins is a mycotoxin produced by *Aspergillus flavus* and *A. parasiticus* fungi found in variety of foods and feeds. In drought stricken region where temperature is very high, aflatoxins are produced by molds in the fields. Aflatoxins are classified as Aflatoxin B_1, Aflatoxin B_2, Aflatoxin G_1, Aflatoxin G_2. Other important group of aflatoxin is Aflatoxin M_1 and Aflatoxin M_2. M series of aflatoxins found in milk and beef from animals that fed on aflatoxin contaminated feed are derived from B series and can be conveniently referred to as hydroxylated Aflatoxin B_1 and Aflatoxin B_2. Aflatoxin (AFB_1) is the most potent hepatocarcinogen of the aflatoxin group.[16,17]

The chemical structure of ochratoxin is shown below. Ochratoxin is produced by several genera *of Aspergillus* and *Penicillium,* which are opportunistic deterioration agent of biological materials. Ochratoxin has been shown to be toxic to animals. Ochratoxins is present generally in few ppb in cereals, coffee and dried fruit.[14] Ochratoxin A is perhaps the most important ochratoxin with wide spread occurrence produced by it is produced by Penicillium verrucosum and P. nordicum.[16] Ochratoxin exists in three forms that are structurally related. Ochratoxin B is a form of Ocratoxin A without chlorination while ochratoxin C is an ethyl ester form of Ochratoxin A. The target organ for ochratoxins is the kidneys and initial interest in this group of toxins was as a cause of Porcine nephropathy [19]. Chemical structure of ochratoxin is shown in figure 3 below.

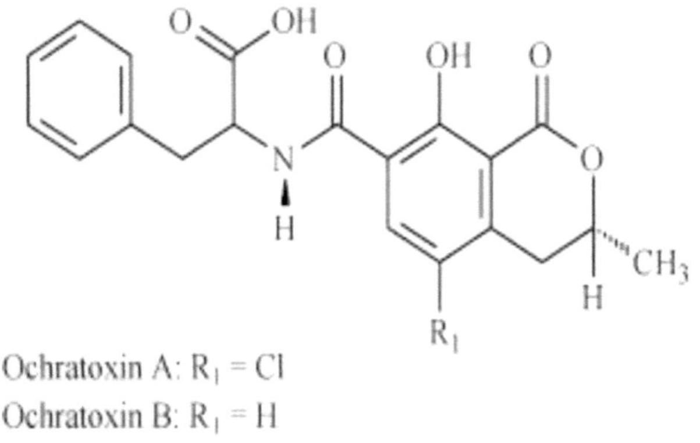

Ochratoxin A: R_1 = Cl

Ochratoxin B: R_1 = H

Figure 3. Chemical structure of Ochratoxin

Penicillium expansum is the mould that produces patulin in rotten tissues of apples. The molecular weight of patulin (4-hydroxy-4-H-furo(3,2-*c*) pyran-2(6H)-one) is 154 and it has the emperical formula of $C_7H_6O_4$. It is an unsaturated lactone with heterocyclic structure. Apart from apples and apple products, which are the food material commonly contaminated with patulin, some other fruits like pears, grapes, peaches also contain patulin in their rotten tissues.[18] The negative effects expressed by paulin include carcinogenicity, mutagenicity, fertility inhibition and teratogenicity[14]

Fumonisin is a group of mycotoxins produced by *Fusarium verticillioides* and commonly contaminate corn and other agricultural produce. About 15 fumonisins have been reported and have been categorized in to four groups ((A, B, C, and P). Fumonisin B1 is the most abundant of all the fumonisins with teratogenic, hepatotoxic and neurotoxic effects. Fumonisin affect sphingolipids metabolism because of similar chemical structure[20]. Most toxic effects of fumonisin are targeted to the brain in man however the target organ in swine is the liver[21].

Bacterial toxins

Bacterial cells produce two types of toxins namely, endotoxins and exotoxins. Many bacterial toxins are proteins coded by bacterial genetic material. However, non- protein toxin (lipopolysaccharide) is found in the cell wall of gram negative bacteria as endotoxins. The different mechanisms of action by which bacterial toxins affect human beings have been explained. For instance, bacterial toxins can either combine with certain structural component or alter the normal functioning of a system. Protein toxins are composed of two moieties; an enzyme moiety and polypeptide part for adsorption to receptors. Furthermore, when bacterial toxins are released in the blood stream, the immune system responds violently to defend the body however this leads to septic shock. *Clostridium botulinum, Salmonella typhimurum* and *E. coli* are known for toxin production. Capillary electrophoresis coupled with laser induced fluorescence can be used for detection and quantification of *Staphylococal* enterotoxin.[22]

Marine toxin

The contamination of seafood with algal toxins lead to different types of poisoning (upon consumption) including paralytic shellfish poisoning (PSP), neurotoxic shellfish poisoning (NSP) and diarrhetic shellfish poisoning (DSP). Paralytic shellfish poisoning is due to the consumption of seafood contaminated with saxitoxins (the most common paralytic shellfish toxin) which adversely affect the peripheral nervous system and there is no antidote. Brevitoxin is another marine toxin that is found in mollusks such as oyster due to accumulation over time however it is lethal to fish. The symptoms of its ingestion include abdominal pain, nausea and vomiting. This condition caused by brevitoxin is referred to as neurotoxic shellfish poisoning. Also, okadaic acid and dinophysistoxins found in mollusks results in diarrhetic shellfish poisoning which is characterizrd by chills and vomiting after 30 minutes to few hours of ingestion. Other marine toxins include tetramine in whelks, trimethylamine oxide and gempylotoxin.[5,22]

Plant toxin

Several toxins in plants (phytotoxins) have been identified. These toxins are most times present in plant in minute amount making their detection difficult and as such sensitive detection technique is required for their detection. Phtytotoxins usually accumulates in different part of plants during growth or storage. Examples include thujone (found in essential oils of plants like sage, clury used for flavoring alcoholic beverages), coumarin, estragole, prussic acid in cherry and apples, cyanogenic glucosides in cassava, tomatine in tomatoes, and oxalic acid in rhubarb plant.[5, 23]

Application of CE to toxin detection in past studies

Several studies have used different modes of capillary electrophoresis coupled to different detection of methods depending on the type of toxin to be detected. Many of the samples were subjected to one form of extraction or the other for optimization of the toxin quantity while some do not follow any extraction step.

Paralytic shellfish toxins (PST) are considered significant health risk; therefore, their detection is important. In the study examined, contaminated mussel was analysed using two different methods. In the first method, Capillary zone electrophoresis (CZE) was coupled with conductivity detector (CD) with counter transient isotachophoresis (tITP) for sensitivity improvement as well as to deal with high conductivity in sample matrix. The only difference in the second method was that UV detector was used to replace conductivity detector. Limit of detection (LOD) established experimentally from the analysis ranged from 74.2 to 1020 ng/ml for tITP–CZE–CD and 14 to 461 ng/ml for tITP–CZE–UV. With these LODs, PST analysis close to the regulatory limit is possible for shellfish. These methods were validated against the official method of PST analysis using HPLC with fluorescence detector. [24]

In another study, mushroom toxins including ibotenic acid, muscimol, and muscarine were separated and detected using capillary electrophoresis coupled with electrospray tandem mass spectrometry (CE-ESI-MS). In developing this method, analytical characteristics including limit of detection, migration time and peak area were evaluated. Migration time was in the range of 0.93% to 1.60% while peak area was in the range of 2.96% to 3.42%. Also, the limit of detection values was at nanomolar concentration level making this method suitable for detection of marine toxins in samples such as human urine containing very minute amount of mushroom toxin.[25]

Patulin in apple juice samples was studied by a developed capillary zone electrophoretic method coupled with UV detector set at 276nm. Optimum conditions including voltage, temperature and injection time were evaluated for the developed method. The method was able to separate and detect patulin in 21 different apple juice samples effectively. The developed method was reproducible with very low limit of detection of 5.9×10^{-3} mu g/ml as well as low limit of quantification (1.79×10^{-2}).[26]

Capillary zone electrophoresis-diode array detection (CZE-DAD) method was developed for the determination of moniliformin, a mycotoxin found in maize. For this method, isolation of moniliformin was done in solid phase extraction column. Sixty-three maize samples were tested for moniliformin and detection sensitivity was comparable to the HPLC-UV method, the official technique of moniliformin detection. The detection limit of CZE-DAD was 0.1 mug MON g^{-1}.[27]

Brevotoxins in fish tissues were analysed in another study using micellar electrokinetic capillary chromatography with laser induced fluorescence detection (MEKC-LIF) with laser excitation at 354nm and fluorescence emission at 410nm. Separation of brevotoxins with MEKC was done with sodium borate/sodium dodecyl sulfate buffer at pH 9.3. Following this, derivatization was done to obtain highly fluorescent product. Detection of brevotoxins at trace level was possible because the limit of detection for the MEKC-LIF method was approximately 4pg/g which is better than other chromatographic method.[28]

Conclusion

With growing interest in fast and accurate instrumental methods of food analysis, capillary electrophoresis presents numerous advantages in food analysis ranging from good separation efficiency using different modes as well as improved sensitivity for detection. Analysis of toxins in food has been done largely with HPLC. However, some of the setbacks experienced with this technology in toxin analysis can be overcome by capillary electrophoresis. Despite the advantages that capillary electrophoresis offers, its use for toxin analysis is comparatively lower, which may be due to few numbers of manufacturers, it is recommended that manufacturers should consider producing this instrument in larger number and food analysts should make do with the advantages capillary electrophoresis offers.

References

1. Cruces-Blanco, C.; Gamiz-Gracia, L.; Garcia-Campana, A.M. Applications of capillary electrophoresis in forensic analytical chemistry. *TrAC, Trends in analytical chemistry (Regular ed.)* **2007,** *26,* 215; 215-226; 226.

2. Anastos, N.; Barnett, N.W.; Lewis, S.W. Capillary electrophoresis for forensic drug analysis: A review. *Talanta* **2005,** *67,* 269-279.

3. Tagliaro, F.; Turrina, S.; Smith, F.P. Capillary electrophoresis: principles and applications in illicit drug analysis. *Forensic Sci. Int.* **1996,** *77,* 211-229.

4. Yin, X.; Wang, E. Capillary electrophoresis coupling with electrochemiluminescence detection: a review. *Anal. Chim. Acta* **2005,** *533,* 113-120.

5. Dolan, L.C.; Matulka, R.A.; Burdock, G.A. Naturally occurring food toxins. *Toxins (Basel)* **2010,** *2,* 2289-2332.

6. Dahlmann, J.; Rühl, A.; Hummert, C.; Liebezeit, G.; Carlsson, P.; Granéli, E. Different methods for toxin analysis in the cyanobacterium Nodularia spumigena (Cyanophyceae). *Toxicon* **2001,** *39,* 1183-1190.

7. Keyon, A.S.; Guijt, R.M.; Gaspar, A.; Kazarian, A.A.; Nesterenko, P.N.; Bolch, C.J.; Breadmore, M.C. Capillary electrophoresis for the analysis of paralytic shellfish poisoning toxins in shellfish: comparison of detection methods. *Electrophoresis* **2014,** *35,* 1496-1503.

8. Li, S.F.Y. *Capillary electrophoresis: principles, practice and applications.* Elsevier: **1992**; Vol. 52,

9. Simó, C.; Barbas, C.; Cifuentes, A. Capillary electrophoresis-mass spectrometry in food analysis. *Electrophoresis* **2005,** *26,* 1306-1318.

10. Xu, Y. Tutorial: capillary electrophoresis. *The Chemical Educator* **1996,** *1,* 1-14.

11. Grossman, P.D.; Colburn, J.C. *Capillary electrophoresis: Theory and practice.* Academic Press: **2012**;

12. Vallejo-Cordoba, B.; Gonzalez-Cordova, A.F. Capillary electrophoresis for the analysis of contaminants in emerging food safety issues and food traceability. *Electrophoresis* **2010**, *31*, 2154-2164.

13. Dabrowski, W.M.; Sikorski, Z.E. *Toxins in food.* CRC Press: **2004**;

14. Turner, N.W.; Subrahmanyam, S.; Piletsky, S.A. Analytical methods for determination of mycotoxins: A review. *Anal. Chim. Acta* **2009**, *632*, 168-180.

15. Zheng, M.Z.; Richard, J.L.; Binder, J. A review of rapid methods for the analysis of mycotoxins. *Mycopathologia* **2006**, *161*, 261-273.

16. Marroquin-Cardona, A.G.; Johnson, N.M.; Phillips, T.D.; Hayes, A.W. Mycotoxins in a changing global environment--a review. *Food Chem. Toxicol.* **2014**, *69*, 220-230.

17. Moss, M.O. Mycotoxin review-1. aspergillus and penicillium. *Mycologist* **2002**, *16*, 116-119.

18. de Souza Sant'Ana, A.; Rosenthal, A.; de Massaguer, P.R. The fate of patulin in apple juice processing: a review. *Food Res. Int.* **2008**, *41*, 441-453.

19. Bayman, P.; Baker, J.L. Ochratoxins: a global perspective. *Mycopathologia* **2006**, *162*, 215-223.

20. Stockmann-Juvala, H.; Savolainen, K. A review of the toxic effects and mechanisms of action of fumonisin B1. *Hum. Exp. Toxicol.* **2008**, *27*, 799-809.

21. Diaz, G.J.; Boermans, H.J. Fumonisin toxicosis in domestic animals: a review. *Vet. Hum. Toxicol.* **1994**, *36*, 548-555.

22. Weinberger, R. Capillary electrophoresis of venoms and toxins. *Electrophoresis* **2001**, *22*, 3639-3647

23. Rietjens, I.M.; Martena, M.J.; Boersma, M.G.; Spiegelenberg, W.; Alink, G.M. Molecular mechanisms of toxicity of important food-borne phytotoxins. *Mol. Nutr. Food Res.* **2005**, *49*, 131-158.

24. Abdul Keyon, A.S.; Guijt, R.M.; Bolch, C.J.S.; Breadmore, M.C. Transient isotachophoresis-capillary zone electrophoresis with contactless conductivity and ultraviolet detection for the analysis of paralytic shellfish toxins in mussel samples. *Journal of chromatography.A* **10,** *1364,* 295; 295-302; 302.

25. Ginterova, P.; Sokolova, B.; Ondra, P.; Znaleziona, J.; Petr, J.; Sevcik, J.; Maier, V. Determination of mushroom toxins ibotenic acid, muscimol and muscarine by capillary electrophoresis coupled with electrospray tandem mass spectrometry. *Talanta* **2014,** *125,* 242-247.

26. Guray, T.; Tuncel, M.; Uysal, U.D. A rapid determination of patulin using capillary zone electrophoresis and its application to analysis of apple juices. *J. Chromatogr. Sci.* **2013,** *51,* 310-317.

27. . Maragos, C.M. Detection of moniliformin in maize using capillary zone electrophoresis. *Food Addit. Contam.* **2004,** *21,* 803-810.

28. Shea, D. Analysis of brevetoxins by micellar electrokinetic capillary chromatography and laser-induced fluorescence detection. *Electrophoresis* **1997,** *18,* 277-283.

YOUR KNOWLEDGE HAS VALUE

- We will publish your bachelor's and master's thesis, essays and papers

- Your own eBook and book - sold worldwide in all relevant shops

- Earn money with each sale

Upload your text at www.GRIN.com and publish for free